Ricardo Uriel Pedrosa
Marcelo Luis Rodrigues

Digestibility of annatto meal with exogenous enzymes

Ricardo Uriel Pedrosa
Marcelo Luis Rodrigues

Digestibility of annatto meal with exogenous enzymes

For Nile tilapia

ScienciaScripts

Imprint

Cover image: www.ingimage.com

This book is a translation from the original published under ISBN 978-613-9-61907-8.

Publisher:
Sciencia Scripts
is a trademark of
Dodo Books Indian Ocean Ltd. and OmniScriptum S.R.L publishing group

120 High Road, East Finchley, London, N2 9ED, United Kingdom
Str. Armeneasca 28/1, office 1, Chisinau MD-2012, Republic of Moldova, Europe
Printed at: see last page
ISBN: 978-620-7-70714-0

SUMMARY

I DEDICATE

To my family, for their unconditional support in my academic life.

ACKNOWLEDGMENTS

First of all to God, for accompanying me at every moment of my life and for guiding me this far.

To my parents, Jorge Curvelo Pedrosa and Marta Vieira Feitosa.

My sister Patricia Vieira Pedrosa.

To my aunt and uncle, Juarez Pedrosa, Marcia Feitosa, Julio Pedrosa and Julia Pedrosa.

To my maternal grandparents Mauro Feitosa and Maria Vieira Batista.

To my advisor, Professor Dr. Marcelo Luis Rodrigues. I am grateful for everything.

The Federal University of Paraiba.

The coordination of the Zootechnics course, all the staff who work there;

To all the teachers from whom I had the opportunity to acquire knowledge inside and outside the classroom, especially Professor Marcia Roseane Targino de Oliviera.

To my friends, Sansao de Paula Homem Neto, Giullyann de Oliveira Salviano, Clesio Morgado de Sousa, Thiago Belo da Silva, Ronaldo, Arivaldo, Leonardo, Eden, Anderson, Adriano Prazeres, Jamille, Messias Nogueira, Tarcisio, Guilherme de Sousa Lima, Emerson Gustavo, Gustavo, Victor Jeronimo, Alexandre Perazzo, Pessoa, Ricardo Martins, Higor Fabio, Anaiane, Luana, Silvaney Santos Araùjo, Tyrone, Talles, Moises, Cintia, Marcelo, Rafael Souza, Hyago, Adler, Felipe and all my friends who were and are present at every achievement in my life.

To the whole group in the fish farming sector at CCA-UFPB, especially Mrs. Lourdinha and Denise.

To Ângela Maria, who was placed on my path by God to make me a better person, with her affection, care and dedication, a very special thank you for being part of my life.

SUMMARY

The aim of this study was to evaluate the addition of an enzyme complex to the digestibility of annatto seed meal for Nile tilapia. A digestibility trial was conducted to evaluate the apparent digestibility coefficients of nutrients and digestible energy of annatto seed meal (FSU) and the effects of supplementing feed with an enzyme complex containing cellulase, protease and amylase on the digestibility of nutrients in juvenile Nile tilapia (*Oreochromis niloticus*). The experimental diets were supplemented with a commercial enzyme complex at levels of 0.0, 0.033, 0.066 and 0.099%. The collection period lasted 15 days and apparent digestibility was determined using the indirect method, using chromium oxide (Cr2O3) as a digestibility indicator. 45 male juvenile nile tilapia weighing approximately 100.00g were randomly distributed in three incubators adapted for digestibility testing. The physical and chemical parameters of the water remained at adequate levels throughout the experiment. The addition of the enzyme complex to the feed improved the apparent digestibility coefficient of dry matter, crude protein and crude energy. Among the levels tested, the 0.033% enzyme complex provided the most significant values.

KEYWORDS: exogenous enzymes, alternative ingredient, *Oreochromis niloticus*

1. INTRODUCTION

The World Health Organization (WHO) recommends annual fish consumption of at least 12 kilos per inhabitant per year, but national consumption is still below this recommendation. However, there was an increase from 6.46 kg to 9.03 kg per inhabitant per year between 2003 and 2009. As a result, the international demand for fish is expected to increase by another 100 million tons per year by 2030, according to the United Nations Food and Agriculture Organization (FAO, 2010).

Brazil is one of the few countries able to meet the growing global demand for fish products, especially through aquaculture, which is an activity capable of transforming unproductive areas of small size or low agricultural yield into highly productive areas (BORGHETTI and OSTRENSKI, 2003).

World production is around 126 million tons. Brazil could become one of the largest producers in the world by 2030, when the country's fish production could reach 20 million tons. The country currently produces approximately 1.25 million tons of fish, 38% of which is farmed (FAO, 2010).

Data from the aquaculture census released by the Ministry of Fisheries and Aquaculture (MPA) in 2010 shows that the Northeast region leads the country in fish production, accounting for a third of the country's (411,400 tons), an increase of 24.10% over the period 2007-2009. Paraiba's fish production increased by 15.22% over the same period, from 11.4 thousand to 13.1 thousand tons a year.

According to the same census, tilapia was the most produced species in 2009, accounting for 39% of all fish from continental fish farming. Worldwide production of tilapia (*Oreochromis niloticus*) was approximately 3.7 million tons in 2010, with Brazil being the largest producer in South America. This is due to the fact that tilapia is one of the most promising species for fish farming due to its rapid growth, ease of obtaining larvae and good fillet yield (FAO, 2010).

The importance of this species is related to its hardiness (HAYASHI et al., 1999), and the ease with which it accepts artificial food at all stages of life (SANTIAGO et al., 1987; MEURER et al., 2002), since in the wild, tilapia nilòtica feeds on the lower trophic levels by ingesting phyto- and zooplankton. In confinement, however, it behaves like an opportunistic species and readily accepts artificial food. This species efficiently uses carbohydrates as an energy source, making it possible to use protein and energy sources of plant origin in the formulation (FURUYA, 2010).

Conventional balanced feeds used in intensive farming are formulated to meet the nutritional needs of fish. However, these feeds have high protein levels and high costs, as they generally use animal products and by-products (LANNA et al., 2004).

The growth of intensive tilapia farming has led to a greater dependence on nutritionally complete balanced feeds, due to the reduced access to natural food in this production condition. In this sense, the search for high-quality feed that maximizes the zootechnical potential of each species is a constant challenge for researchers. The intermittent quest to reduce feed costs must be seen as a set of integrated measures involving research and extension institutions, feed companies and producers.

The use of plant-based foods in diets aims to reduce feed costs and consequently reduce production costs, so the replacement of certain products and by-products from the agro-industry that are used as ingredients in fish diets with substitute products has been studied by several authors and shows effective results in terms of nutrient utilization for certain species (PEZZATO et al., 2001).

Fish feed formulation is mainly based on corn, soybean meal and fishmeal, which often make fish production unfeasible (SANTOS et al., 2009), due to the great variability in price, dependence on supply throughout the year and the difficulty of transportation to regions that do not produce these foods.

According to EL-SAYED, (1999), KUBITZA, (1997) and PEZZATO et al., (2000), feed costs in intensive and semi-intensive fish farming generally exceed 50% of the cost of production, which reinforces research into alternative feeds that has been growing in recent years, with the priority of maintaining or improving fish performance.

One viable way of reducing the cost of feed would be to use by-products and co-products from the agro-industry. In the case of the regions furthest from the feed-producing centers, such as the North and Northeast, the use of alternative regional foods is an opportunity to reduce the cost of feed. These regions traditionally buy conventional food and feed from producing regions (the South and Southeast), which increases the cost of production (HISANO and PORTZ, 2007).

In order to use agro-industrial residues in balanced diets for fish, it is necessary to know their nutritional composition and, therefore, to assess their digestibility. The lack of data on the digestibility of ingredients has been a major problem in the manufacture of these diets.

According to PEZZATO et al. (2004), knowledge of the digestibility of agro-industrial by-products has made it possible to use various foods in complete fish diets. Studies show that foods with similar chemical compositions can present differences in the evaluation of the digestibility coefficient.

However, there is little information on the effect of exogenous enzymes on the digestibility of alternative foods used in fish feed. The aim of this study was to evaluate different levels of enzyme complex on the digestibility of annatto meal for Nile tilapia.

2. LITERATURE REVIEW

2.1. Species under study: Nile tilapia (*Oreochromis niloticus)*

Today, after the common carp (*Cyprinus carpio),* tilapia is the most cultivated species in the world, because it is resistant to handling, has very tasty meat and is extremely resistant to adverse environmental conditions and diseases (PROENÇA and BITTENCOURT, 1994). It is a fish species with an omnivorous eating habit and is very versatile, as it adapts both to extensive cultivation without any technology being used, and to the system of rearing in net tanks with complete rations and high production technology (MEURER et al., 2002).

It stands out for its hardiness and rapid growth in intensive farming (HAYASHI et al., 1999), for the ease with which it obtains larvae, for its precocity, for its good organoleptic characteristics with low fat content (0.9 $g.100g^{-1}$ of meat) and calories (117 $kcal.100g^{-1}$ of meat). It feeds on the basic items in the trophic chain and accepts a wide variety of foods. It responds with the same efficiency to ingesting proteins of plant and animal origin, and has the physiological capacity to adapt to different environments and production systems, it can withstand high stocking densities and low levels of dissolved oxygen, high fillet yields (35 to 40%) and the absence of intramuscular spines in its fillet, which makes it viable to industrialize it for sale on the domestic and foreign markets (HILDSORF, 1995). It also has lower protein requirements compared to carnivorous species, as it is at the bottom of the food chain (CHELLAPPA et al., 1996; FITZSIMMONS, 2000).

It is a tropical species whose ideal temperature for growth varies between 25 and 30°C. Its growth is affected below 15°C and it cannot withstand temperatures of around 9°C (CASTAGNOLLI, 1992; KUBITZA, 2000; GONZALEZ and QUEVEDO, 2001; ONO and KUBITZA, 2003; CYRINO and CONTE, 2006).

The name tilapia comes from a group of fish comprising around 70 species, divided into four genera, all belonging to the family Cichlidae and subfamily Tilapiinae. Its origin is African, with a natural geographical distribution area that stretches from East Africa (Nile Basin, Congo) to West Africa (Niger and Senegal River Basins), and research into the breeding of this species began in the Belgian Congo (now Zaire) at the beginning of the 19th century. From 1924 onwards, its breeding was intensified in Kenya and its expansion to other parts of the world began in Malaysia (CAMPO, 2008).

This species has been spread by humans in Israel, Southwest Asia, the United States and South America (SIQUEIRA FILHA et al. 1999). In Brazil, it was introduced by the Secretary of Agriculture of the State of São Paulo in 1952 to contain the proliferation of algae and aquatic macrophytes in dams (OSTRENSKY et al. 2008). And in 1971, through the National Department for Works Against Drought (DNOCS) in the dams of the Northeast, spreading to the whole country (PROENÇA and BITTENCOURT, 1994; CASTAGNOLLI, 1996).

In terms of digestibility studies, tilapia have stood out in research involving the digestibility of energy and nutrients from conventional and alternative sources of plant origin (FAGBENRO, 1998; PEZZATO et al.., 2001), since they tolerate diets with high levels of carbohydrates (DEGANI and REVACH, 1991; NRC, 1993; STICKNEY, 1997), a fact attributed to their morphological and physiological adaptations, which include pharyngeal teeth, low stomach pH (<1.5) and long intestines (KUBARIK, 1997; HANLEY; 1987). These findings were reinforced by DEGANI et al. (1997), in studies carried out on nile tilapia and hybrid tilapia (*Oreochromis aureus* x *O. niloticus*).

2.2. Alternative foods

One viable way of reducing the cost of feed would be to use by-products and co products from the agro-industry. In the case of the regions furthest from the feed-producing centers, the use of alternative regional foods would be an interesting opportunity to reduce the price of feed (TACHIBANA, 2007). These regions traditionally buy conventional food and feed from producing regions, which increases the cost of production.

Replacing the ingredients usually used in fish feed with certain products and by-products from the agro-industry, crop residues and products not intended for human consumption has been presented as an alternative economic practice.

Alternative feeds generally replace standard feeds. The formulation of fish feed is mainly based on corn, soybean meal and fishmeal, which, due to the great variability in price and depending on supply throughout the year and the difficulty of transportation to regions that do not produce these feeds, often make fish production impractical (RIBEIRO et al, 2005).

According to HAYASHI et al. (1999), more than 90% of the feed used is vegetable grain or by-products, with corn being the energy source and soybean meal the protein source most commonly used in the formulation of omnivorous fish feed. These products can be totally or partially replaced by alternative feeds to reduce production costs, especially in the off-season.

Alternative feeds generally replace standard feeds. In the different regions of Brazil, there are various foods with the potential to be used as alternative feeds for Nile tilapia (RIBEIRO et al, 2005).

According to PEZZATO et al. (2001), replacing certain agro-industrial products and by-products used as ingredients in fish feed with substitute products has become an alternative economic practice.

Theoretically, feed is made up of protein and energy foods, which do not have very flexible prices. As a result, every time the price of a staple food such as corn or soy rises, so does the price of the substitute food (PEZZATO et al., 2009).

Thus, studies aimed at understanding alternative foods, products or co-products that can partially replace traditional ingredients in the diets of non-ruminant animals are essential.

In order to reduce production costs, a large number of experiments have been carried out at research institutions in Brazil testing unconventional or alternative feeds for poultry, pigs and fish. It is first necessary to research the nutritional information of the foods (chemical composition, energy, digestibility, restrictions, anti-nutritional factors, etc.), so that nutritionists can include them in databases and formulate commercial rations at minimum cost (ROSTAGNO et al., 2008).

Among the alternative foods of plant origin most investigated by researchers are: sorghum, triticale, millet, algaroba, annatto and tomato residue meal as energy sources and leucaena leaf or seed meal, palm oil cake, canola meal, cotton, dehydrated alcohol yeast and cassava as a protein source. Coconut bran, on the other hand, can be used both as an energy source and as a protein source (SANTOS et al., 2009).

Given the possibility of using a variety of alternative foods, knowledge of the digestibility coefficient of foods and nutrients allows for the formulation of rations that best meet nutritional requirements, avoiding both physiological and environmental overload, thus increasingly bringing the biological optimum closer to the economic optimum (PEZZATO, 2009).

2.3. Urucum

The annatto tree (*Bixa orellana*) is a tree that can reach up to 6 meters in height. They are trees of the bixaceae family native to tropical America with large light green leaves and pink flowers. The fruits are capsules with malleable spines, which turn red when ripe. They then open to reveal small reddish seeds, arranged in series, surrounded by red aril and their seeds are commonly used as a natural dye.

In Brazil, annatto dye powder is known as *paprika and* is used in cooking to enhance the color of food. Annatto has been the dye most used by the national industry, representing around 90% of the natural dyes used in Brazil and 70% worldwide (TOCCHINI and MERCADANTE, 2001). The main applications of annatto-based dyes in the food industry are: in the sausage sector, in the dairy industry (Prato cheese, butter), in the ice cream and confectionery industries and it is estimated that more than 2.8 tons per year are consumed in other foods and in other non-food applications (cosmetics and pharmaceuticals) (FRANCO, 2002).

Brazilian annatto production in 2002 was 11,582 tons, with the state of São Paulo being the largest producer with 2,058 tons, followed by Bahia with 1,991 tons, Rondônia with 1,757 tons, Parà with 1,498 tons and Paranà with 1,058 tons (IBGE, 2004). Of this total, 60% is destined for the manufacture of colorifico, 30% for the manufacture of dyes and 10% for export.

Paraiba has already topped the national ranking with 48% of production (IBGE, 1989). In 2001, this state implemented the Urucuzeiro Revitalization Program, achieving an increase in planted area of over 70%. The goals set by the program refer to the introduction of seedlings with high genetic potential, which will provide an 80% increase in productivity in the state of Paraíba between the 5th and 6th year, when the crop stabilizes (SILVA et al., 2008), benefiting around 400 rural producers in 23 municipalities: Guarabira, Alagoinha, Araçagi, Cuitegi,

Pirpirituba, Piloes, Piloezinhos, Alagoa Grande, Areia, Bananeiras, Solânea, Serraria,

Borborema, Duas Estradas, Sertaozinho, Itapororoca, Mamanguape, Alhandra, Mataraca, Serra da Raiz, Lagoa de Dentro, Sao Sebastiao de Lagoa de Roça and Lagoa Seca, located in the geographic micro-regions of Guarabira, Brejo paraibano and Litoral paraibano.

Currently, the cultivation of the annatto tree is concentrated mainly in the municipalities of Araçagi, Areia, Belém, Alagoinha, Pirpirituba, Duas Estradas, Alagoa Grande, Serra da Raiz, Lagoa de Dentro and Guarabira, which is the polarizing center for all commercialization. Considered a backyard crop, the vast majority of growers are small producers with land modules of up to 20 hectares.

The main chemical constituents of annatto include saturated and unsaturated fatty acids, sugars, calcium, cellulose, iron, phospholipids, phosphorus, orelin, potassium, proteins, saponins, tannins, vitamins A, B2 and C. The main carotenoid components of annatto are bixin, methyl bixin, nor bixin, trans bixin, β carotene, lutein, cryptoxanthin and zeaxanthin. There are basically two compounds: bixin and norbixin (or orelin). Bixin is orange-red in color, soluble in alcohol, oils and fats and insoluble in water. Norbixin, which is yellow in color, is soluble in water, chloroform, alkaline solutions and vegetable and animal oils. The composition of the aril (the substance that coats the annatto seed), from which the coloring substances are extracted, is quite complex (CARVALHO, 1990; AMBROSIO et al., 2006; MELENDEZ, 2004).

In an experiment to determine the apparent digestibility coefficients, energy and protein of annatto meal in growing pigs (UTIYAMA, 2001), it was determined that the digestible energy and protein were 2365 kcal/kg and 8.80% respectively, and that in relation to a diet based on corn and soybean meal, annatto meal can be included up to a level of 10% without influencing the performance of growing pigs.

Determining the bromatological composition of this by-product from the extraction of

Urucum pigments, WURTS and TORREBLANCA (1983) found values of 13.70% Crude Protein, 14.40% Crude Fiber, 1.4% Ethereal Extract and 5.9% Mineral Matter. Work carried out by OLIVEIRA (2004) showed similar results: Crude Protein 10.8% and 11.50%, Crude Fiber 12.6%, Ethereal Extract 4.8% and Ash 4.6%. PEZZATO et al. (2004) reports approximate values for Crude Protein (11.5%), Ethereal Extract (5.0%) and Mineral Matter (5.0%).

After bixin extraction, the annatto by-product has the following bromatological composition: 85.08% dry matter, 14.57% crude protein (CP), 55.91% neutral detergent fiber (NDF), 23.39% acid detergent fiber (ADF), 2.90% ethereal extract (EE) and 32.52% hemicellulose, based on DM (GONÇALVES et al., 2004).

For feed manufacturing, this can be an alternative and economically viable source of feed supplementation, making it a profitable source for the industries that obtain it as a by-product.

2.4. Digestibility

Chemical analysis and feeding tests are the first steps in determining the nutritional value of a foodstuff (PEZZATO, 2004). However, after ingestion, the effective assimilation of the food depends on the use that the animal organism is able to make of it (MAYNARD and LOOSLY, 1966). Animal species assimilate food differently and this variation is quantified by determining their digestibility coefficients (ANDRIGUETO et al., 1982). According to these authors, the digestibility of a feed is defined as the animal's ability to digest and absorb the nutrients and energy contained in it.

The growing development of tilapia farming in Brazil and around the world has brought with it the need to formulate diets that enable greater production efficiency. In order to produce diets with these characteristics, it is necessary to know the digestibility of the nutrients and energy of the main ingredients that make up a diet, as well as the nutritional requirements for the species and life stage that we wish to produce (FURUYA, 2000). In this way, tilapia have stood out in studies involving the digestibility of energy and nutrients from alternative sources of animal origin (SANTOS et al., 2009).

Determining digestibility has been one of the main tools for assessing the quality of a diet or ingredient, indicating its nutritional value, as well as the levels of undigested nutrients, which will make up most of the waste accumulated in the aquatic environment (FURUYA et al., 2001). Thus, the more digestible a nutrient is, the greater its availability will be.

According to PEZZATO et al. (2002), various methodologies and structures have been used to determine the digestibility coefficients of ingredients and fish feed. Among the most commonly used methods are: a) a collection tank equipped with a valve for concentrating feces, b) extrusion of the final contents of the intestine, c) dissection of the distal portion of the intestine, d) mechanical anal suction and e) capture of the feces released into the tank. In an analysis of these methods of determining the apparent digestibility coefficient, the same author concludes that in the distal portion of the intestine there is a significant process of absorption of the diet, therefore, the dissection method underestimates the digestibility of the material collected, so the digestibility coefficients measured by the indirect method, with feces collected in the collection tanks, are more reliable.

MORALES et al. (1999) state that the determination of the food consumed and the total collection of feces is made difficult by the aquatic environment, which is why the indirect method of determining digestibility has been used in preference. According to ZIMMERMANN and JOST (1998), the direct method is subject to errors and it is difficult to determine precisely how much has been ingested and excreted. Using the indirect method eliminates the need to collect all the excreta and allows the fish to eat at will (NRC, 1993).

The indirect method of collecting feces consists of a cylindrical fiberglass tank with a conical bottom, in which the feces that decant in the tank are conducted through external pipes, where they are deposited in tubes attached to the bottom (SAKOMURA and ROSTAGNO, 2007). This method uses a marker that is included in the diet in concentrations of 0.5 to 1.0%, which is then evaluated in the feces. The most commonly used marker is chromium oxide (Cr2O3), but other markers can also be used (NRC, 1993), but for an indicator to be considered ideal, it must be completely indigestible and non-absorbable, insoluble in water, have no pharmacological action and pass evenly through the digestive tract, have easy and rapid chemical determination and preferably be a natural constituent of the diet (GODDARD and MCLEAN, 2001).

Using chromium oxide, the apparent digestibility, according to NOSE (1966), is estimated using the following equation:

$$CDA = 100 - \left[100.\left(\frac{\%I_r}{\%I_f}\right).\left(\frac{\%N_f}{\%N_r}\right)\right]$$

In which:

ADC = apparent digestibility coefficient (%);

$\%I_r$ and $\%I_f$ = % Indicator in diet and feces, respectively;

$\%N_f$ and $\%\ N_r$ = % of nutrient in feces and diet, respectively.

The partial or total replacement of some of the ingredients usually used in fish feed factories with certain agro-industrial residues, i.e. products not intended for human consumption, has been presented as an alternative economic practice. The digestibility of these alternative feeds has been studied by various authors and several have shown effective results in this area of animal nutrition knowledge (SILVA, 2011).

2.5. Exogenous enzymes

In fish nutrition studies, knowledge of food digestibility coefficients alone is not enough; more analysis is needed to increase these coefficients through proper food processing, including those that are difficult to digest, as occurs with some species when fed diets rich in carbohydrates and proteins of plant origin (SILVA et al., 2007).

According to the author, today, through genetic recombination and mutation techniques, biotechnology has enabled the industrial production of specific exogenous enzymes, using various types of fungi, bacteria and plants, and has brought benefits to animal farming. The use of exogenous digestive enzymes has attracted attention for their inclusion in aquaculture feed. Industrial enzymes can be purchased on the Brazilian market and have been introduced as a supplement in animal feed.

There are various types of enzymes with different substrates of action, as well as enzyme complexes made up of different enzymes. The main functions of supplementation with exogenous enzymes in the diets of monogastric animals include: the removal or destruction of anti-nutritional factors present in foods of plant origin, an increase in the digestibility of diets, the potentiation of the action of endogenous enzymes and the reduction of environmental pollution caused by the excretion of nutrients (GARCIA et al., 2000).

According to SIGNOR (2008), exogenous enzymes increase the digestibility of nutrients, improving the productive performance of fish and reducing nutrient excretion. They can also reduce feed costs due to their low purchase price and because they are included in small quantities in diets. Enzymes have also been used with the aim of incorporating lower quality raw materials into the feed of domestic animals (NERY et al., 2000) and fish (NG et al.,

2002), reducing the impact caused by waste and taking advantage of ingredients available in each region.

GUIMARAES et al. (2009) reports that in order to achieve high production efficiency at low costs, it is necessary to improve the use of diets and increase the variety of ingredients that can be used in formulations. Feed producers, like fish producers, are highly dependent on the cost of ingredients, as well as their nutritional value. In this sense, research has sought to solve these problems by studying the nutritional value of ingredients, as well as food additives that can improve the nutritional value of these ingredients or their diets, while also reducing the eutrophication of the environment. It can be assumed that enzymes help the digestibility of nutrients and reduce the elimination of excreta into the water (MORA-JAIMES et al., 2002).

Most enzymes are globular proteins, with a tertiary or quaternary structure, which act as biological catalysts, increasing the speed of chemical reactions in the body, without themselves being altered in the process (CHAMPE and HARVEY, 1989). They are highly specific for the substrates and direct all metabolic events. Digestive enzymes have an active site which allows them to act by breaking a particular chemical bond under favorable conditions of temperature, pH and humidity (PENZ JÙNIOR, 1998). Enzyme complexes act by breaking down cell walls (TORRES et al., 2003).

The plant-based ingredients most commonly used in feed for non-ruminant animals are relatively rich in starch and protein, but contain non-digestible components present in the cell wall, such as non-starch polysaccharides (NSPs), oligosaccharides and other non-carbohydrates (glycoproteins, phenolic esters, lignin) (SIGNOR, 2008). OLIVEIRA et al. (2007) describe that PNAs exert a barrier effect to the action of hydrolytic enzymes by trapping starch molecules and/or other nutrients (such as cell wall proteins) inside endosperm cells and that heat treatments are not able to fully release non-digestible nutrients from plant-based ingredients. The covalent bonds between PNAs and lignin limit the digestibility of polysaccharides when ingested by non-ruminant animals (FISCHER et al., 2002).

Evaluating the use of an enzyme complex containing cellulase, protease and amylase on the digestibility of juvenile Nile tilapia (*Oreochromis niloticus*) at levels of 0.0, 0.025, 0.050, 0.075 and 1.0% inclusion, OLIVEIRA et al. (2007) observed a quadratic effect on the apparent digestibility coefficients of dry matter, crude protein and crude energy, with values of 0.0495; 0.051 and 0.0498%, respectively, with the best results for the 0.05% level. This same trend was confirmed by OGUNKOYA et al. (2005) who observed positive effects on the apparent digestibility coefficient of dry matter, crude protein, lipids, phosphorus and energy when they evaluated the incorporation of soybean meal and an enzyme complex consisting of xylanase, amylase, cellulase, protease and β-glucanase in the diet of rainbow trout *Oncorhynchus mykiss*. However, STONE et al. (2003) evaluated the enzyme complex composed of β-glucanase and xylanase, which act specifically on non-starch polysaccharides, in feed for silver perch *Bidyanus bidyanus* and reported that they observed no difference in the dry matter, protein and energy of the feed.

The choice of cellulase enzyme for enzyme complexes can be a major factor in obtaining positive results for the apparent digestibility of dry matter and crude protein, especially crude energy (OLIVEIRA et al., 2007). Cellulase stands out as the most important of the group of enzymes, as it degrades plant cell walls (SAHA et al., 2006).

The fiber present in food is mainly composed of cellulose, a glucose polymer with a β -1,4 glycosidic bond (SAHA et al., 2006). Possibly, the action of the cellulase enzyme on cellulose provides a greater energy contribution from crude fiber in relation to the total energy of the diet, as well as a shorter retention time of the food bolus in the digestive tract, which could have a positive impact on fish performance (OLIVEIRA et al., 2007).

Research involving exogenous enzymes in fish diets has shown that the digestibility coefficients show positive and increasing responses up to a certain level of inclusion, after which the values stabilize or decline (OLIVEIRA et al., 2007). It is therefore important to determine the ideal level of inclusion of each enzyme compound available on the market in fish feed.

NUNES et al. (2006) evaluated the inclusion of an enzyme complex consisting of amylase, lipase and protease on the performance of tambaqui *Colossoma macropomum* and reported that the exogenous enzymes amylase and lipase influenced zootechnical performance at inclusion levels of 0.05 and 0.2%, respectively, while the same did not occur for protease.

However, for enzymes to be used in non-ruminant feed, they must resist and retain their activity after the manufacturing and digestion processes, as the molecular structure of enzymes is quite fragile and can therefore be denatured by heat, alkalis, heavy metals and other oxidizing agents (GRAHAM and INBORR, 1991).

DIAS et al. (2002) evaluated protease over different periods (0; 0.5; 1; 5; 10 and 15 minutes) of incubation at a temperature of 80°C and treatments with different pH (5.0; 2.3 without pepsin and 2.3 with pepsin). According to the same author, there was no influence from the time of exposure to temperature, demonstrating that this enzyme remains active after the pelletizing process and only lowering the pH was enough to partially inactivate it, since only 64.7% of the enzyme retained its original activity.

In view of the above, it is important to develop an appropriate processing methodology and/or use feed additives that maximize the availability of nutrients offered to fish, thus improving growth, reducing farming time and increasing aquaculture production. The use of nutrients in fish feed reduces the excretion of nutrients into the aquatic environment, thus reducing environmental pollution and consequently increasing the carrying capacity of current production systems (SIGNOR, 2008).

3. MATERIAL AND METHODS

The trial was conducted in the Fish Farming Sector of the Department of Zootechnics, on Campus II of the Federal University of Paraiba, using 60 juvenile Nile Tilapia (*Oreochromis. niloticus*) of the GIFT strain, sexually reversed in the initial phase, with a live weight of approximately 100.0±10.0g.

The animals were distributed in four water tanks with a capacity of 350 liters, equipped with heaters and aerated. In each water tank, the fish were kept in circular net tanks made from plastic mesh, each containing 15 fish. These were used to house the fish and facilitate feeding and feces collection, according to PEZZATO et al. (2002).

The fish were kept in water tanks throughout the day and fed at will in small fractions from 8:00 am to 5:00 pm, by means of manual feeding. After the last portion of food, the fish were removed to fiberglass conical-bottomed incubators (Figure 1) with a capacity of 180 liters, adapted for collecting feces the following day.

Figure 1 - Conical-bottom fiberglass incubators adapted for collecting feces.

The apparent digestibility of the feed was determined according to the NRC (1993) by the indirect method of collecting feces, using 0.2% chromium oxide ($Cr\ O_{23}$) as an external indicator, a practical diet as a reference (Table 1) and three test diets containing 0.033;

0.066 or 0.099% of the enzyme complex (Allzyme SSF®), consisting of dry yeast, amylase, protease and cellulase, supplied by Alltec do Brasil Agroindustrial Ltda. The diets consisted of 70% of the reference diet and 30% of annatto meal as an alternative regional source to be tested with the addition of the enzyme complex at the respective inclusion level according to the treatment.

Table 1 shows the percentage composition of the foods used in the reference diet for determining the apparent digestibility coefficient in tilapia.

Table 1. Percentage composition of the reference diet used to determine the apparent digestibility coefficients of food for Nile tilapia.

Ingredients	**%**
Corn	58,34
Corn gluten	17,26
Soybean meal	17,17
Soybean oil	2,70
Bi-calcium phosphate	2,67
L-Lysine HCL	0,90
Common salt	0,50
Calcario	0,15
L-Threonine	0,12
L-Tryptophan	0,08
DL-Methionine	0,05
Mineral and vitamin premix[1]	0,50
BHT[2]	0,01
3	0,2

Chromium Oxide	
Total	100,00
Calculated Nutrient Composition	
Digestible energy (kcal/kg)	3.200
Crude protein	27
Dry matter	91,71
Crude fiber	2,33
Càlcio	0,8
Available phosphorus	0,6
Total lysine	1,5
Total Met+Cis	0,92
Met.total	0,51
Total Threonine	0,98

1 Guarantee levels per kilogram of product (supieinais): vit. A, 1,200,000ui; vit. D3, 200,000ui; Vit. E, 12,000mg; Vit. K3, 2,400mg; Vit. B1, 4,800mg; Vit. B2, 4,800mg; Vit. B6, 4,000mg; Vit. B12, 4,800mg; Folic Acid, 1,200mg; Pantothenate Ca, 12.000mg; Vit. C, 48.000mg; Biotin, 48mg; Choline, 65.000mg; Niacin, 24.000mg; Iron, 10.000mg; Copper, 6.000mg; Manganese, 4.000mg; Zinc, 6.000mg; Iodine, 20mg; Cobalt, 2mg; Selenium, 20mg;[2] Butyl-Hydroxy-Toluene (antioxidant);[3] Cr2O3 (indicator).

The alternative feeds purchased were analyzed together with the other ingredients in the basal feed. Analyses of dry matter (DM), crude energy (CE), crude fiber (CF), crude protein (CP) and ash (CZ) were carried out at the Animal Nutrition Laboratory of the Agricultural Sciences Center of the Federal University of Paraiba, according to the methodology of SILVA and QUEIROZ (2002). Chromium oxide readings were taken at the Animal Nutrition Laboratory of the Federal University of Minas Gerais. The chemical analyses were determined according to the recommendations of AOAC (1995) and the evaluation of chromic oxide (Cr O_{23}) as an indicator was carried out using the method described by (SILVA, 1990).

To prepare the diet, all the food was ground in a knife mill and a 0.5 mm diameter sieve, then the ingredients were weighed and homogenized, chromium oxide was added, then they were pulverized with water (60° C) in the proportion of 12% of their total weight and then aggregated to a diameter of 8 mm in a meat grinder and dehydrated in a forced ventilation oven (55° C) for a period of 24 hours.

The water in the boxes and incubators was kept under constant aeration using a central blower with a porous stone attached to keep the dissolved oxygen constant. The water temperature was kept constant using two 300W thermostats. The physical and chemical variables of the water (pH, conductivity, dissolved oxygen and temperature) in the boxes and incubators were monitored daily in the morning and afternoon. The water in the feces collection tank was renewed daily at 8:30 am and 5:30 pm using water from a 5000L reservoir.

After seven days of acclimatization to the local conditions and adaptation to the test rations, the first collection of faeces was carried out, totaling fifteen days of faeces samples collected, which were stored at -4°C for later analysis. The material collected was then dehydrated in a forced ventilation oven at a temperature of 55° C for 24 hours. After drying, the material was ground in a ball mill, identified and stored in a refrigerator for later analysis.

The fish were fed every two hours, from 8:00 am to 5:00 pm, by hand until they were apparently satiated, at which point they were no longer able to catch the granules provided as they moved from the surface to the bottom of the water tanks.

The apparent digestibility coefficients of dry matter, crude protein and energy in the diet were determined according to the expression described by Nose (1960). The apparent digestibility coefficients of energy and nutrients in the feed were calculated according to the formula described by CHO and SLINGER (1979):

$$CDA = \frac{CDA_{DT} - CDA_{DR} \cdot x}{y}$$

In which:

CDA = coefficient of apparent digestibility of energy or nutrients;

CDA^{DT} = *apparent digestibility coefficient of the energy or nutrients in the test diet;*

CDADR = coefficient of apparent digestibility of energy or nutrients in the reference diet;

x = proportion of the referenced diet;

y = proportion of the test ingredient.

4. RESULTS AND DISCUSSION

The physico-chemical parameters of the water in the feed tanks, measured in the morning, showed values within those required for tilapia cultivation according to KUBITZA (2000), so they possibly had no influence on the results of the apparent digestibility coefficients of the experimental diets. The average temperature, pH, ammonia and dissolved oxygen values of the water in the boxes were 26.6°C, 6.77, 0.008mg/L and 5.8mg/L respectively.

The chemical compound values of annatto seed meal are shown in Table 2.

Table 2. Chemical composition in Dry Matter (DM), Crude Protein (CP), Crude Fiber (CF), Ether Extract (EE), Ash (CE) and Crude Energy (CE) of annatto seed meal (USF).

Ingredients	MS	PB	FB	EE	CZ	EB
			%			kcal/kg
FSU	93,89	11,90	41,86	10,05	5,55	5.146

Values expressed as 100% of Dry Matter

UTIYAMA (2001), working with annatto seeds in pig feed, found a value higher than this study for crude protein (14.73), and lower for dry matter (86.71) and crude energy (3,743 kcal). However, GARCIA et al. (2009) found similar values for crude protein (12.99) and dry matter (91.32) and lower values for crude fiber (2.34) and ether extract (5.0).

The bromatological composition varied little in the studies that used the seed *in natura, as* well as in the studies that used the processed annatto seed residue.

Evaluating the nutritional composition of annatto seed residue, SILVA et al. (2005) found values close to those of the annatto seed meal in this study for: crude protein (12.5%) and crude energy (4,400 kcal).

The average values (%) of the apparent digestibility coefficients (CDa) of dry matter (DM), crude protein (CP) and apparent digestible energy (EDa) of annatto seed meal (FSU) and the reference breed are shown in Table 4.

Table 3. Values of the apparent digestibility coefficients of dry matter (CDaMS), crude protein (CDaPB) and digestive energy of the food evaluated.

Enzyme complex level	CDa%		
	MS[1]	PB[2]	ED[3]
0,000%	41,26	75,82	55,62
0,033%	76,04	91,81	78,40
0,066%	12,54	72,18	14,05
0,099%	24,72	71,98	38,74

The apparent digestibility coefficients of DM, CP and ED (Table 3) of the feed containing 0.033% inclusion of the enzyme complex tested were higher than the other inclusion levels tested. The values (%) of the digestibility coefficients found in this study are therefore excellent, since digestibility studies with tilapia usually use diets made up of plant-based foods with low digestibility.

The chemical compound values of the feeds used are shown in Table 4.

Table 4. Analyzed chemical coniposity of the experimental rations with different levels of inclusion of the enzyme complex.

Compound	Enzyme complex (%)			
	0,0	0,033	0,066	0,099
Dry matter (%)	96,17	91,82	93,22	91,73
Crude protein (%)	22,50	27,13	27,0	29,99
Crude energy kcal/kg	4.639	4.517	4.463	4.824

Ether extract (%)	9,02	8,09	7,03	8,57
Crude fiber (%)	5,21	8,4	13,05	9,6
Ash (%)	8,95	5,08	4,94	4,97

The interference of crude fiber (13.05%) observed in the diet with 0.066% inclusion of the enzyme complex (Table 4) on the apparent digestibility of DM, CP and ED is due to the fact that it alters the rate of gastric emptying, by acting on the activity of digestive enzymes, by capturing lipid micelles, and thanks to its interaction with the surface of the intestinal wall, it interferes with the absorption of nutrients (MADAR and THORME, 1987).

The diet with the addition of 0.033% of the enzyme complex may have increased the CDa of PB, DM and ED, as it provided a greater contact surface between the substrate and the enzyme and consequently improved the digestion of carbohydrates and lipids associated with the protein in the feed.

OLIVEIRA et al. (2007) concluded that the inclusion of an enzyme complex composed of cellulase, protease and amylase (Allzyme Vegpro-Alltech) at levels of 0; 0.025; 0.050; 0.075 and 0.1% in diets formulated with ingredients of plant origin for tilapia had a quadratic effect on the apparent digestibility coefficient of dry matter, crude protein and crude energy, corresponding to 0.049; 0.052 and 0.049% inclusion, respectively, and that the apparent digestibility coefficient of starch, calcium and phosphorus increased with the inclusion of 0.05%.

The results found are comparable to those reported by SILVA et al. (2007), who evaluated the effect of the apparent digestibility of nutrients and energy in feed supplemented with exogenous digestive enzymes for juvenile tambaqui, where they found that among the four levels of inclusion tested (0.0, 0.05, 0.10 and 0.15%), the level of inclusion of 0.05% contributed to the increase in digestibility. This author also stated that the presence of cellulase in the enzyme complex may have been a factor in the results achieved, since they can be explained by taking into account that this enzyme has an action in the degradation of the constituents of the cell walls present in the ingredients of the feed based on vegetable products, which, together with xylanases and glucanases, make up the main enzymes for the degradation of NAPs (Non-starch Polysaccharides), which are not synthesized by non-ruminants, making the monomers of carbohydrates, proteins and lipids in plant cells bioavailable to fish.

The results observed in this study differ from those of GUIMARÂES et al. (2009), who, working with increasing levels of enzyme complex inclusion, found that the ADC of the CP ration increased with the addition of the product tested. GARCIA et al. (2000), however, did not observe any effect from the addition of enzyme complexes to corn and soy-based diets for broiler chickens.

According to OLIVEIRA (2007), in most studies involving the addition of exogenous enzymes to fish diets, the apparent digestibility coefficients show positive responses up to a certain level of inclusion, showing a tendency to stabilize or decline, which was observed in the results obtained at the 0.066% and 0.099% levels in this study. The same author evaluated the effects of supplementing feed with an enzyme complex containing cellulase, protease and amylase on the digestibility of nutrients in juvenile Nile tilapia (*Oreochromis niloticus*). The author concluded that the addition of the enzyme complex to the feed, composed of soybean meal and corn, improves the apparent digestibility coefficient of dry matter, crude protein, crude energy, starch, calcium and phosphorus, at a level of 0.05%, which corroborates the results presented.

According to GUIMARAES et al. (2009) the supplementation of digestive enzymes does not always provide a positive response. For the enzyme to work, there needs to be a specific substrate in the diet, the correct dosage of enzymes and the ability of the enzymes to overcome the barriers present in the stomach - low pH and the action of proteolytic enzymes such as pepsin - and the temperature to which the feed is subjected during pelleting.

OGUNKOYA et al. (2005) observed the influence of an enzyme complex composed of amylase, xylanase, protease, cellulase and α-galactosidase in diets for trout *Oncorhynchus mykiss,* which increased the apparent digestibility coefficients of dry matter, crude protein, lipids, phosphorus and energy. STONE et al. (2003), however, did not observe this same effect when supplementing diets with α-galactosides and xylanase for *silver perch* (*Bidyanus bidyanus*).

In the majority of studies involving the addition of exogenous enzymes to fish diets, the apparent digestibility coefficients showed positive and increasing responses up to a certain level of inclusion and, from that point onwards, showed a tendency to stabilize or decline. A similar behavior was observed in this study, so excessive enzyme contraction may have a negative influence on nutrient digestibility.

5. CONCLUSION

The enzyme complex consisting of amylase, protease and cellulase, when added to the feed of juvenile tilapia at a level of 0.033%, improves the apparent digestibility of nutrients and digestible energy of the feed containing 30% annatto seed meal.

6. REFERENCES

AMBRÓSIO, C.L.B, CAMPOS, F.L.B, FARO, Z,P. Caratenoids as an alternative to hypovitaminosis A. Revista Nutriçâo. PUCCAMP, v.12, p.233-243p, 2006.

ANDRIGUETO, J.M.; PERLY, L.; MINARDI, I. et al. A. Nutriçâo Animal. Vol. 1, Ed. Universidade do Paranà-PR, Nobel, 395p, 1982.

ASSOCIATION OF OFFICIAL ANALYTICAL CHEMISTS (AOAC). Pepsin digestibility of animal protein feeds. In: OFFICIAL METHODS OF ANALYSIS OF AOAC INTERNATIONAL, Virginia. Ed. Patricia Cunniff, chap. 4. p.15-16, 1995.

BORGHETTI, N.R.B.; OSTRENSKY, A.; BORGUETTI, J.R. *Aqüicultura - uma visco geral sobre a produção de organismos aquâticos no Brasil e no mundo*. Curitiba: integrated group of aquaculture and environmental studies. Botucatu, SP. *Anais...* Botucatu: SBZ, 1998. p.165-178, 2003.

CAMPO, L.F.C. LA TILAPIA ROJA: una evolucion de 26 anos, de la incertidumbre al exito. Mexico, 2008. 147p.

CARVALHO ,P.R.N,. Extraction and use of annatto dye. In: Sâo José, A,R., REBOUÇAS, T.N.H. A cultura do urucum no Brasil. Vitória da Conquista BA, UESB, p.69-76, 1990.

CASTAGNOLLI, N. Breeding freshwater fish. Jaboticabal: FUNEP. 1992.189p.

CASTAGNOLLI, N. Aquaculture for the year 2000. Brasilia: CNPq. p. 95, 1996.

CHAMPE, P.C.; HARVEY, R.A. Enzymes. In: *Bioquimica ilustrada*. 2.ed. Sâo Paulo: Artes Médicas, 1989. p.53-66.

CHELLAPPA, N. T.; CHELLAPPA, S.;CAMPERO, D. C. F. The feeding habits and food types of the nilotic tilapia, *Oreochromis niloticus*. In: SIMPÒSIO BRASILEIRO DE

AQUICULTURA, 9., 1996, Piracicaba. *Abstracts*... Piracicaba: ABRAq, 1996. p.106.

CHO, S.Y.; SLINGER, S.J. Apparent digestibility measurements in feedstffs for rainbow trout. In: HALVER, J.E.; TIEWS, K. (Ed.). Finfish Nutrition and Fishfeed Technology. Berlin: Heenemann, p. 234-247, 1979.

CYRINO, J.E.; CONTE, L.; Caged tilapia: production and economy. In: José Eurico Possebon Cyrino and Elisabeth Criscuolo Urbinati (Eds.). AquaCiência 2004: Special Topics in Aquatic Biology and Aquaculture. Jaboticabal: Brazilian Society of Aquaculture and Aquatic Biology, ch.12, p.151-171, 2006.

DEGANI, G.; REVACH, A. Digestive capabilities of three commensal fish species: carp, Cyprinus carpio L., tilapia, Oreochromis aureus x O. niloticus, and African catfish, Clarias gariepinus (Burchell, 1882). Aquacult. Fish. Manage., Amsterdam, v. 22, p. 397-403, 1991.

DEGANI, G., VIOLA, S., YEHUDA, Y. 1997. Apparent digestibility of protein and carbohydrate in feed ingredients for adult tilapia (*Oreochromis aureus x O. niloticus*). Israeli J. Aquac, v.49, p.115-123, 1997

DIAS, J.C.C.A. et al. Evaluation of the *in vitro* stability of a commercial protease. Arquivo Brasileiro de Medicina Veterinària e Zootecnia, v.54, n.6, p.618-622, 2002.

EL-SAYED, A.F.M. Alternative dietary protein sources for farmed tilapia, *Oreochromis* spp. Aquaculture, v.179, p.149-168, 1999.

FAGBENRO, O.A. Apparent digestibility of various legume seed meals in Nile tilapia diets, Aquac. Intern, v.6, p.83-87,1998.

FAO - Food and Agriculture Organization of the United Nations. Aquaculture statistics 2010. Available at:< <http://www.fao.com.br/ statistica>. Accessed on September 21, 2010.

FITZSIMMONS, K. Tilapia: most important aquaculture species of the 21st century. In:

PROCEEDINGS FROM THE FIFTH INTERNATIONAL SYMPOSIUM ON TILAPIA AQUACULTURE, 2000, Rio de Janeiro. *Proceedings*... Rio de Janeiro: ISTA, 2000. p.3-8.

FISCHER, G. et al. Performance of broilers fed diets based on corn and soybean meal, with or without the addition of enzymes. Revista Brasileira de Zootecnia, v.31, n.1, p.402-410, 2002.

FRANCO, C. F. O.; SILVA, F. C.; FILHO, J. C.; NETO, M. B.; SÂO JOSÉ, A. R.; REBOUÇAS, T. N. H.; FONTINELLI, I. E. C. Urucuzeiro agribusiness of natural dyes. Joâo Pessoa: EMEPA-PB, 2002. 120 p.

FURUYA, W.M. Apparent digestibility of amino acids and substitution of fishmeal protein for soybean meal protein based on the ideal protein concept in diets for Nile Tilapia (*Oreochromis niloticus*). 2000. Faculty of Veterinary Medicine and Zootechny - Universidade Estadual Paulista, Botucatu, 2000.

FURUYA, W. M.; PEZZATO, L. E.; MIRANDA, E. C.; FURUYA, V. R. B.; BARROS, M. M. Apparent digestibility coefficients of energy and nutrients of some ingredients by nile tilapia, *Oreochromis niloticus* (L.) (Thai strain). Acta Scientiarum Maringà, v. 23, n. 2, p. 465-469, 2001

FURUYA, W.M. Tabelas Brasileiras para a Nutriçâo de tilâpias. p.19, 2010.

GARCIA, E.R.M.; MURAKAMI, A.E.; BRANCO, A.F. et al. Effect of enzyme supplementation in rations with soybean meal and extruded whole soybeans on nutrient digestibility, nutrient flow in ileal digesta and broiler performance. *Rev. Bras. Zootec.*, v. 29, p.1414-1426, 2000.

GARCIA,E.A et al. Performance and egg quality of conventional layers fed annatto seeds (Bixa orellana L.) ground in the diet. Vet e Zootecnia, p.689-697, v.16, n.4, 2009.

GODDARD, J.S.; McLEAN, E. Acid-insoluble ash as an inert reference material for digestibility studies in tilapia *(Oreonchromis aureus)*. Aquaculture, v. 194, p. 93-98, 2001.

GONÇALVES, J.S.; NEIVA, J.N.M.; SÂ, C.R.L. et al. Nutritional value of elephant grass silage *(Pennisetum purpureum* Schum.) cv. Roxo with different levels of addition of annatto seed by-product (*Bixa orellana* L.). In: ANNUAL MEETING OF THE BRAZILIAN SOCIETY OF ZOOTECHNICS, 41, Campo Grande, 2004. Proceedings... Campo Grande: SBZ, 2004.

GONZÂLEZ, C.E.; QUEVEDO, E.T. Cultivo de las tilâpias roja (Oreochromis spp.) y plateada (*Oreochromis niloticus*), chap.XIII. p. 283-299. GOMEZ, H.R.; DAZA, P.V.; AVILA, M.C.C. Fundamentos de Acuicultura Continental. Bogotá: Instituto Nacional de Pesca y Acuicultura, 2001, 423p.

GRAHAM, H.; INBORR, J. Enzymes in monogastric feeding. Agro Industry Hi-tech, v. 2, n. 1, p. 45-48, 1991.

GUIMARAES, I.G. et al. Apparent digestibility of feeds containing enzyme complex for nile tilapia. *Arq. Bras. Med. Vet. Zootec.* vol.61, n.6, pp. 1397-1402, 2009.

HANLEY, F. The digestibility of foodstuffs in the effects of feeding selectivity on digestibility determination in tilapia, *Oreochromis niloticus* (L.). Aquaculture, v.66, p.163-179, 1987.

HAYASHI, C.; BOSCOLO, W.R.; SOARES, C.M.; MEURER, F. Use of different degrees of grinding of ingredients in feed for Nile tilapia (*Oreochromis niloticus,* L.) in the growth phase. Acta Scientiarum. 1999. 733- 737p.

HILDSORF, A.W.S. Genetics and culture of red tilapia, a review. Boletim do Instituto de Pesca, v.22, p.73-87, 1995.

HISANO, H.; PORTZ, L. *Reducing feed costs for tilapia*: the importance of protein. Bahia Agric., v.8, n. 1, nov. 2007

Brazilian Institute of Geography and Statistics (IBGE). Municipal Agricultural Production. Available at: < http://www.ibge.gov.br/home/estatistica/economia/pam/2002/default.shtm > Accessed on: May 10, 2012.

KUBARIK, J. Tilapia on highly flexible diets. Feed International, v.6, p.16-18, 1997.

KUBITZA. F. *Fish nutrition and feeding*. Piracicaba- SP 74p (1997).

KUBITZA, F. Tilapia: Technology and planning in commercial production. 1[a] ed. Jundiai: Fernando Kubitza, 2000. 289p.

LANNA, E. A.T.; PEZZATO, L. E.; FURUYA, W. M.; VICENTINI, C. A.; CECON, P. R.; BARROS, M. M. Crude fiber and oil in practical diets for Nile tilapia (*Oreochromis niloticus*) fry. Revista Brasileira de Zootecnia, v.33, n.6, p.2177-2185, 2004.

MADAR, Z.; THORNE, R. Dietary fiber. Progress in Food and Nutrition Science, v. 11, n. 2, p. 153-174, 1987.

MAYNARD, L.A.; LOOSLY, J.K. *Nutriçao Animal*. Rio de Janeiro: McGraw Hill, 1966.

MELENDEZ, A.J, Nutritional importance of carotenoid pigments. Arch Latino Nutr. v.54, 149-155p, 2004.

MEURER, F.; HAYASHI, C.; BOSCOLO, W.R; Soares, C.M. Lipídeos na alimentação de alevinos revertidos de tilâpia do Nilo (*Oreochromis niloticus*). Revista Brasileira de Zootecnia. 31(2): 566 - 573, 2002.

MINISTRY OF FISHERIES AND AQUACULTURE (MPA). Statistical bulletin of fishing and aquaculture in Brazil 2008-2009. Available at:

<http://www.sepaq.pa.gov.br/files/u1/anuario da pesca completo.pdf>, accessed on 08/05/2012.

MORA-JAiMES, G.; BÂCENA-GAMA, R.; MENDOZAMARTiNEZ, G.D.; GONZALES-MUNOZ, S.S.; HERRERAHARO, J.G. Respuesta productiva y fermentación ruminal em borregos alimentados con grano de sorgo tratado con amilasas. Agrociencia, v.36, p.31-39, 2002.

MORALES, A.E.; CARDENETE, G.; SANZ, A. et al. Re-evaluation of crude fiber and acid-insoluble ash as intermarkers, alternative to chromic oxide, in digestibility studies with rainbow trout (*Oncorhynchus mykiss*). Aquaculture, v.179, p.71-79, 1999.

NERY, V.L.H.; LIMA, J.A. F.; NELO, R.C.A. Addition of exogenous enzymes for piglets from 10 to 30 kg of weight. Rev. Bras. De Zootec., 29(3):794-802, 2000.

NG, W-K.; LIM, H-A.; LIM, S-L.; IBRAHIM, C-O. Nutritive value of palm kernel meal pretreated with enzyme or fermented with *Trichoderma koningii* (Oudemans) as a dietary ingredient for red hybrid tilapia (*Oreochromis* sp.). *Aquaculture Research,* 33:1199-1207, 2002.

NOSE, T. On the digestion of food protein by gold-fish (*Carassius auratus* L.) and rainbow trout (*Salmo irideusG.*).Bulletin of Freshwater Fisheries Research Laboratory, v.10, p.11-22, 1960.

NOSE, T. Recent advances in the study of fish digestion in Japan. In: SIMPOSIuM ON FINFISH NuTRITION AND FISH FEED TECHNOLOGY, Belgrade: EIFAC/FAO, 15p.1966.

NATIONAL RESEARCH COUNCIL - NRC. Nutrients requirements of fish. Washington

D. C.: National Academy of Science, 114 p. 1993

NUNES, E.S.S. et al. Exogenous digestive enzymes in the feeding of juvenile tambaqui. Pesquisa Agropecuâria Brasileira, v.41, n.1, p.139-143, 2006.

OGUNKOYA, A.E. et al. Dietary incorporation of soybean meal and exogenous enzyme cocktail can affect physical characteristics of faecal material egested by rainbow trout (*Oncorhyncus mykiss*). Aquaculture, v.254, p.466-475, 2005.

OLIVEIRA, N.T. E. Metabolizable energy of food and egg and meat quality of Japanese quails fed diets containing annatto colorific and supplementary niacin. 2004. 47 p. Thesis (Doctorate in Animal Production) - Universidade Estadual do Norte Fluminense, Campos dos Goytacazes, 2004.

OLIVEIRA, G.R. et al. Digestibility of nutrients in multi-enzyme complex feeds for Nile tilapia. Revista Brasileira dc Zootecnia, v.36, n.6, p. 1945- 1952, 2007.

ONO, E. A.; KUBITZA, F. Fish farming in net tanks. 3ª ed. Jundiai: Eduardo A. Ono, 2003. 112p.

OSTRENSKY, A.; BORGHETTI, J. R.; SOTO, D. Aqüicultura no Brasil o desafio é crescer. Brasilia, 2008. p. 276.

PENZ JÛNIOR, A.M. Enzymes in poultry and pig feed. *In*: SIMPÒSIO DE ADITIVOS NA PRODUÇÃO DE RUMINANTES E NÃO RUMINANTES, 1998,

PEZZATO, L.E.; MIRANDA, E.C.; BARROS, M.M.; PINTO, L.G.Q.; PEZZATO, A.; FURUYA, W.M., Nutritional value of coconut bran for Nile tilapia (*Oreochromis niloticus*). *Acta Scient. Anim. Sci.* 22 (3): 695-69, 2000.

PEZZATO, L.E.; BARROS, M.M.; FRACALOSSI, D.M. et al. Apparent digestibility of ingredients by Nile tilapia (*Orcochromis niloticus*). Revista Brasileira de Zootecnia, Viçosa,

2001.

PEZZATO et al. Evaluation of two methods for determining apparent digestibility coefficients with Nile tilapia (Oreochromis niloticus). Acta Scient. Anim. Sci. v. (3): 695-69, 2002.

PEZZATO, L. E.; MIRANDA, E. C.; BARROS, M. M.; FURUYA, W. M.; PINTO, L. G. Q. Apparent digestibility of dry matter and crude protein and digestible energy of some alternative feeds by Nile tilapia (Oreochromis niloticus). Revista Acta Scientiarum Animal Sciences. Maringà, v.26, n.3, p.329-337, 2004.

PEZZATO, L.E., BARROS, M.M., FURUYA, W.M.,. Nutritional value of feeds in the formulation of rations for tropical fish. Revista Brasileira de Zootecnia, 38, 43-51, 2009.

PROENÇA, C.E.M.; BITTENCOURT, P.R.L. Manual de Piscicultura Tropical. Brasilia: IBAMA, 1994.196p.

RIBEIRO, P.A.P.; GOMIERO, J.S.G.; LOGATO, P.V.R. Manejo alimentar de peixes. Lavras: Nùcleo de Estudos em Aquacultura, 2005. v.1, p.1-13. (Technical Bulletin).

ROSTAGNO, H. S.; NASCIMENTO, A. H.; ALBINO, L. F. T.; RODRIGUES, P. B. Retrospective and Challenges of Animal Production - Poultry and Pigs. DZO/U.F.V., Viçosa - MG, 2008.

SAHA, S. et al. Characterizations of cellulase-producing bacteria from the digestive tract of tilapia, *Oreochromis mossambica* (Peters) and grass carp, *Ctenopharyngodon idella* (Valenciennes). Aquaculture Research, v.37, p.380-388, 2006.

SAKOMURA,N.K., ROSTAGNO,H.S., Métodos de pesquisa em nutriçâo de monogàstricos. Jaboticabal: Funep, chap.2, p. 79, 2007.

SANTIAGO, C.B.; ALDABA, M.B.; REYES, O.F. Influence of feeder rate and diet form on growth and survival of Nile tilapia (*Oreochromis niloticus*). Aquaculture, 64(2):277-282, 1987.

SANTOS, E. L. et al. APPARENT DIGESTIBILITY OF COCONUT FLOUR AND GOIABA WASTE BY NILE TILANY (*Oreochromis niloticus).* REVISTA CAATINGA - UNIVERSIDADE FEDERAL RURAL DO SEMI-ÂRIDO (UFERSA. v.22, n.2, p.175-180, April/June 2009

SIGNOR, A. A. FEED PROCESSING AND THE USE OF MULTIENZYME COMPLEX IN THE PRODUCTION OF NILE TILANY (*Oreochromis niloticus*) YOUNG. Dissertation (Master's Degree in Animal Science) - Federal University of Paraiba, Marechal Càndido Rondon Campus, 2008.

SILVA, D.J. Food analysis (chemical and biological methods). 2.ed. Viçosa, MG: Universidade Federal de Viçosa, 166p.1990.

SILVA, D.J.; QUEIROZ, A.C. Food analysis: chemical and biological methods. Viçosa: UFV, 2002. 235p.

SILVA, J. H. V. da; Effects of the Inclusion of Urucum Seed Residue (*Bixa Orellana* L.) in the Diet for Broiler Chickens: Performance and Carcass Characteristics. R. Bras. Zootec., v.34, n.5, p.1606-1613, 2005

SILVA, M. C. D.; BOTELHO, J. R.; SOUZA, A. G. Kinetic study of bixin dye by dynamic thermal decomposition. Tecnol. & Ciên. Agropec., João Pessoa, v.2., n.1, p.11-14, mar. 2008

SILVA, D. A. Digestibility and performance of nile tilapia fry fed diets containing different levels of annatto seed and forage palm bran. Dissertation (Master's Degree) - Federal

University of Paraiba, Areia, 2011

SILVA, J. A. M da.; PEREIRA-FILHO, M.; CAVERO, B. A. S., OLIVEIRA- PEREIRA, M.I. Apparent digestibility of nutrients and energy in feed supplemented with exogenous digestive enzymes for juvenile tambaqui (Colosssoma macropomum Cuvier, 1818): Acta Amazonica . vol. 37(1) 2007: 157 - 164

SIQUEIRA FILHA, N. T.; SIQUEIRA, A. T.; LIRA, J. M. T.; SANTOS, A. J. G. Sexual reversal of nilotic tilapia (*Oreochromis niloticus*) in green water, with larvae from artificial incubation. In: BRAZILIAN CONGRESS OF FISHING ENGINEERING, 11., 1999, Recife. Proceedings... Recife: FAEP-BR, 1999. v. 1, p. 147-157.

STICKNEY, R.R. Tilapia nutrition, feeds and feeding. In: COSTA-PIERCE, B.A., RAKOCKY, J.E. (Eds.) Tilapia aquaculture in the Americas. v.1, WAS, Baton Rouge. p.34-54, 1997.

STONE, D.A.J. et al. Carbohydrate utilization by juvenile silver perch *Bidyanus bidyanus* (Mitchell). II Digestibility and utilization of starch and its breakdown products. Aquaculture Research, v.34, n.2, p.109-121, 2003.

TACHIBANA, L. Triticale in Nile tilapia feed. 54f. Thesis (Doctorate in Aquaculture) - Universidade Estadual Paulista, Centro de Aquicultura, Jaboticabal-SP, 2007.

TOCCHINI, L.; MERCADANTE, A.Z. Extraction and determination by CLAE of bixin and norbixin in colorificos. Ciência e Tecnologia de Alimentos, Campinas, v.21, p.310313, 2001.

TORRES, D.M.; COTTA, J.T.B.; TEIXEIRA, A.S. et al. Diets based on corn and soybean meal supplemented with enzymes in broiler feeding. Ciência Agrotecnológica, v.27, n.1, p.199-205, 2003.

UTIYAMA,C.E. Utilization of the residue of processed urum seeds (Bixa orellana L.) in the feed of growing pigs.Piracicaba, Escola Superior de Agricultura de Auqueiroz. USP, p.43 2001, Master's dissertation.

WURTS, M. L.; TORREBLANCA, R. A. Anàlisis de la semilla *Bixa orellana* L. (Achiote) y del desecho generado en la extracción de sus pigmentos. Archivos Latinoamericanos de Nutrición, v.33, p.606-619, 1983.

ZIMMERMANN, S.; JOST, H.C. Recent advances in fish nutrition: phased nutrition in intensive fish farming. In; SYMPOSIUM ON FISH MANAGEMENT AND NUTRITION, 1., 1998, Piracicaba. Anais...Piracicaba: Conselho Brasileiro de Nutriçâo Animal,. p.123-162, 1998.

MIX
Papier aus verantwortungsvollen Quellen
Paper from responsible sources
FSC® C105338

Printed by Books on Demand GmbH, Norderstedt / Germany